## Student Activity Manual
## for

# *MANUFACTURING*
## *SYSTEMS*

by
**R. Thomas Wright**
Professor, Industry and Technology
Ball State University
Muncie, Indiana

Publisher
**The Goodheart-Willcox Company, Inc.**
Tinley Park, Illinois

# INTRODUCTION

You are about to enter an exciting study of manufacturing. It is the source of much of the human-made world—the technology that makes life better for all of us.

Are you ready to apply your knowledge and skill in challenging, new activities? If so, the *Manufacturing Systems Student Activity Manual* is for you! The chapters in this student activity manual are divided into sections that correspond to the sections in the text. The flexible design of the student activity manual will allow you to use it as you perform a variety of experiments and manufacture a variety of products. You will not be restricted to a few structured activities. Instead, you, your class, and your teacher can work together in selecting what you will manufacture during your study of manufacturing systems.

As you work through the activities, you will soon realize that there is generally more than one right answer to design problems and process selection. This student activity manual gives you and your teacher the freedom to select and design the products you will produce. However, with freedom comes responsibility. You will be responsible for using your creativity and intelligence to contribute to class discussions and decisions. You will also be responsible for maintaining records and notes of class decisions. Additionally, you will be responsible for carefully observing demonstrations and recording procedures for completing major processes. Finally, at ALL times, you will be responsible for working safely.

It is wise to set goals. During this class, you will be challenged to accept excellence as your goal. You will be able to select and use manufacturing products more intelligently. You will also become acquainted with many types of manufacturing careers.

# CONTENTS

# Section 1
# MANUFACTURING SYSTEMS

**Activity 1-1: Line Production**

**Chapter 1**

Name _____

Date _____

Score _____

Period _____

You and your teacher will select a product to be manufactured in class using line production. On the grid below, sketch the product. Then, on the following page, complete the procedure/safety chart.

Name of the product to be manufactured: _____

Sketch of the product:

*(Continued)*

Name _____

Job on the manufacturing line: _____

As your teacher demonstrates your job on the manufacturing line, list the procedures you are to use and record the safety precautions that you are to follow.

| Procedure | Safety precaution |
|---|---|
| 1. | |
| 2. | |
| 3. | |
| 4. | |
| 5. | |
| 6. | |
| 7. | |
| 8. | |
| 9. | |
| 10. | |
| 11. | |
| 12. | |
| 13. | |
| 14. | |
| 15. | |

## Activity 1-2: Manufacturing System Resources     Chapter 2

Name _____     Score _____

Date _____     Period _____

List the inputs, processes, and outputs that are used for the part of a manufacturing system shown in each photo.

1.

**Automotive part manufacture:**

Inputs: _____
_____
_____
_____
_____
_____

Processes: _____
_____

Outputs: _____
_____

2.

**Automatic part manufacture:**

Inputs: _____
_____
_____
_____
_____
_____

Processes: _____
_____

Outputs: _____
_____

*(Continued)*

**Aircraft product design:**

3.

Inputs: _____
_____
_____
_____
_____
_____

Processes: _____
_____

Outputs: _____
_____

**Product quality improvement meeting:**

4.

Inputs: _____
_____
_____
_____
_____
_____

Processes: _____
_____

Outputs: _____
_____

(Photos courtesy of General Motors Corp.)

## Activity 1-3: Using Manufacturing System Resources

Name _____   Score _____

Date _____   Period _____

Name of the product that was manufactured:_____

List the manufacturing system inputs that were used to produce the product your class manufactured as part of this unit of study.

| Inputs | |
|---|---|
| Materials: | |
| Human labor (types): | |
| Information: | |
| Capital (tools and equipment): | |
| Energy (types): | |
| Finances (product cost): | |

*(Continued)*

Name _____

Describe the manufacturing system processes that were used to produce the product your class manufactured.

| Processes | |
|---|---|
| Transformation: | |
| Management: | |

List the manufacturing system outputs that resulted from the product your class manufactured.

| Outputs | |
|---|---|
| Desirable: | |
| Undesirable: | |

# Section 2
# MANUFACTURING MATERIALS

**Activity 2-1: Identifying Manufacturing Materials**                **Chapters 3, 4, and 5**

Name _____          Score _____

Date _____          Period _____

Your teacher will give you some manufacturing materials to identify. You are to approach the task as a material scientist would. First, list as many properties and clues as you can that will help you identify the material. Then use these properties to help you make an informed "guess" as to the correct name of the material.

Sample #1:

    Type: ☐ Wood      ☐ Metal      ☐ Plastic      ☐ Ceramic      ☐ Composite

    Color: _____

    Relative weight: ☐ Light      ☐ Average      ☐ Heavy

    Hardness: ☐ Soft      ☐ Average      ☐ Hard

    Flexibility: ☐ High (flexible)      ☐ Average      ☐ Low (stiff)

    Material name: _____

Sample #2:

    Type: ☐ Wood      ☐ Metal      ☐ Plastic      ☐ Ceramic      ☐ Composite

    Color: _____

    Relative weight: ☐ Light      ☐ Average      ☐ Heavy

    Hardness: ☐ Soft      ☐ Average      ☐ Hard

    Flexibility: ☐ High (flexible)      ☐ Average      ☐ Low (stiff)

    Material name: _____

Sample #3:

    Type: ☐ Wood      ☐ Metal      ☐ Plastic      ☐ Ceramic      ☐ Composite

    Color: _____

    Relative weight: ☐ Light      ☐ Average      ☐ Heavy

    Hardness: ☐ Soft      ☐ Average      ☐ Hard

    Flexibility: ☐ High (flexible)      ☐ Average      ☐ Low (stiff)

    Material name: _____

**Activity 2-2: Manufacturing Materials:**     **Chapters 3, 4, and 5**
                 **Grades and Sizes**

Name _____     Score _____

Date _____     Period _____

Use a catalog to gather the following information for each material assigned your group.

Material: _____

     Does the material come in standard sizes? ☐ Yes     ☐ No

     Sizes available: _____

     Grades available: _____

     How is the material sold (unit of measure)? _____

     What is the price per unit measure? _____

Material: _____

     Does the material come in standard sizes? ☐ Yes     ☐ No

     Sizes available: _____

     Grades available: _____

     How is the material sold (unit of measure)? _____

     What is the price per unit measure? _____

Material: _____

     Does the material come in standard sizes? ☐ Yes     ☐ No

     Sizes available: _____

     Grades available: _____

     How is the material sold (unit of measure)? _____

     What is the price per unit measure? _____

**Activity 2-3: Testing Manufacturing Materials**     **Chapters 3, 4, and 5**

Name _____     Score _____

Date _____     Period _____

Obtain material testing directions from your teacher. Record the data requested on this page and the following page. Then draw conclusions about the manufacturing material(s) you tested.

Material: _____

Property tested: _____

Sketch the test specimen here:

*(Continued)*

Name _____

As your teacher demonstrates the test, list the procedures you are to use and record the safety precautions that you are to follow.

| | Test procedure used | Safety precaution |
|---|---|---|
| 1. | | |
| 2. | | |
| 3. | | |
| 4. | | |
| 5. | | |
| 6. | | |
| 7. | | |
| 8. | | |
| 9. | | |
| 10. | | |

Describe your findings.

_____

_____

_____

Analyze your results.

_____

_____

_____

_____

Name three products in which this product should be used.

_____

_____

_____

# Section 3
# MANUFACTURING PROCESSES

**Activity 3-1: Types of Processes**        **Chapter 6**

Name _____        Score _____

Date _____        Period _____

You and your teacher will select a product to be manufactured in class. On the grid below, sketch the product. Then, on the following pages, complete the related charts and answer the following questions.

Name of the product to be manufactured: _____

Sketch of the product:

*(Continued)*

In the chart below, list the primary processes used to produce the industrial materials that were the material inputs for the product. Then list the secondary processes used to make the product.

| Primary processes |
| --- |
| 1. |
| 2. |
| 3. |
| 4. |
| 5. |
| 6. |
| 7. |
| 8. |
| 9. |
| 10. |

| Secondary processes |
| --- |
| 1. |
| 2. |
| 3. |
| 4. |
| 5. |
| 6. |
| 7. |
| 8. |
| 9. |
| 10. |

*(Continued)*

**Activity 3-1, Chapter 6**
**(Continued)**

Name _____

As your teacher demonstrates the processes, list the procedures you are to use and record the safety precautions that you are to follow.

| Procedure | Safety precaution |
|---|---|
| 1. | |
| 2. | |
| 3. | |
| 4. | |
| 5. | |
| 6. | |
| 7. | |
| 8. | |
| 9. | |
| 10. | |
| 11. | |
| 12. | |
| 13. | |
| 14. | |
| 15. | |
| 16. | |
| 17. | |
| 18. | |
| 19. | |
| 20. | |

*(Continued)*

**Activity 3-1, Chapter 6**
**(Continued)**                          Name _____

Analyze the product that was produced by answering the following questions.

1. How would you evaluate this product in terms of quality?
   _____ Excellent          _____ Good          _____ Fair          _____ Poor

   Give reasons for your evaluation.

   _____

   _____

   _____

   _____

   _____

   _____

   _____

   _____

   _____

   _____

   _____

   _____

2. How would you build the product differently using the knowledge you gained in building
   the first product prototype?

   _____

   _____

   _____

   _____

   _____

   _____

   _____

   _____

   _____

   _____

   _____

   _____

   _____

## Activity 3-2: Casting and Molding Processes

**Chapter 7**

Name _____

Date _____

Score _____

Period _____

You and your teacher will select a part or product to be manufactured in class. On the grid below, sketch the part or product. Then, on the following pages, complete the related charts.

Name the part or product to be manufactured: _____

Sketch of the part or product:

*(Continued)*

Name _____

Casting or molding process to be used:_____

As your teacher demonstrates the process, list the procedures you are to use and record the safety precautions that you are to follow.

| Procedure | Safety precaution |
|---|---|
| 1. | |
| 2. | |
| 3. | |
| 4. | |
| 5. | |
| 6. | |
| 7. | |
| 8. | |
| 9. | |
| 10. | |
| 11. | |
| 12. | |
| 13. | |
| 14. | |
| 15. | |
| 16. | |
| 17. | |
| 18. | |
| 19. | |
| 20. | |

*(Continued)*

# Activity 3-2, Chapter 7
# (Continued)

Name _____

Mark the appropriate box of each major step in producing a casting. Then write a brief description that describes each step you used in producing the product.

| | |
|---|---|
| 1. Preparing a mold: ☐ Expendable (one-shot) | ☐ Permanent |

| | | |
|---|---|---|
| 2. Preparing material: ☐ Melt | ☐ Dissolve | ☐ Compound |

| | |
|---|---|
| 3. Introducing materials into the mold: ☐ Pour | ☐ Force |

| | | |
|---|---|---|
| 4. Solidifying the material: ☐ Cool | ☐ Dry (remove solvent) | ☐ Chemical action |

| | |
|---|---|
| 5. Extracting the casting or molded part: ☐ Destroy | ☐ Open |

Describe and explain any defects in your casting.

| Defect | Possible cause |
|---|---|
| | |

*(Continued)*

Name _____

Select three casting processes you and your classmates completed. List their proper names in the boxes across the top of the chart. For each process, write a brief description about the five basic concepts listed on the left of the chart.

| | | | |
|---|---|---|---|
| Mold | | | |
| Preparing the material | | | |
| Introducing the material into the mold | | | |
| Solidifying the material | | | |
| Extracting the casting | | | |

## Activity 3-3: Forming Processes

**Chapter 8**

Name _____

Score _____

Date _____

Period _____

You and your teacher will select a part or product to be manufactured in class. On the grid below, sketch the part or product. Then, on the following pages, complete the related charts.

Name of the part or product to be manufactured: _____

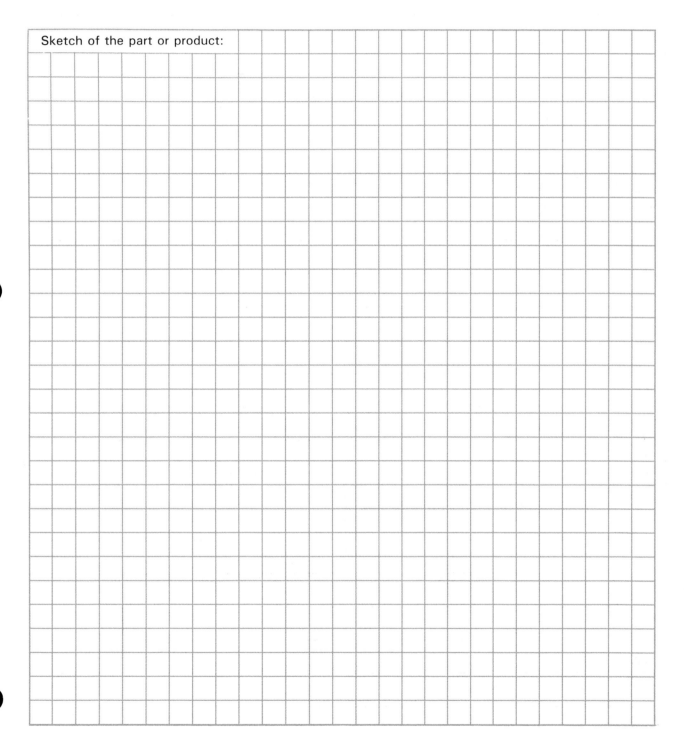

Sketch of the part or product:

*(Continued)*

# Activity 3-3, Chapter 8 (Continued)

Name _____

Forming process to be used: _____

As your teacher demonstrates the process, list the procedures you are to use and record the safety precautions that you are to follow.

| Procedure | Safety precaution |
|---|---|
| 1. | |
| 2. | |
| 3. | |
| 4. | |
| 5. | |
| 6. | |
| 7. | |
| 8. | |
| 9. | |
| 10. | |
| 11. | |
| 12. | |
| 13. | |
| 14. | |
| 15. | |
| 16. | |
| 17. | |
| 18. | |
| 19. | |
| 20. | |

*(Continued)*

Name _____

Mark the appropriate box of each major step in producing a formed part. Then write a brief description that describes each step you used in producing the product.

| | | |
|---|---|---|
| 1. Shaping device: ☐ Die | | ☐ Roll |
| | | |
| | | |
| | | |
| 2. Material temperature: ☐ Cold | | ☐ Hot |
| | | |
| | | |
| | | |
| 3. Source of forming pressure: ☐ Machine tool | | ☐ Other |
| | | |
| | | |
| | | |

Describe and explain any defects in your formed part.

| Defect | Possible cause |
|---|---|
| | |
| | |
| | |
| | |
| | |

*(Continued)*

**Activity 3-3, Chapter 8**
**(Continued)**

Name _____

Select three forming processes you and your classmates completed. List their proper names in the boxes across the top of the chart. For each process, write a brief description about the three basic concepts listed on the left of the chart.

| | | | |
|---|---|---|---|
| Shaping device | | | |
| Material temperature | | | |
| Source of forming pressure | | | |

## Activity 3-4: Separating Processes

Name _____

Date _____

Score _____

Period _____

You and your teacher will select a part or product to be manufactured in class. On the grid below, sketch the part or product. Then, on the following pages, complete the related charts.

Name of the part or product to be manufactured: _____

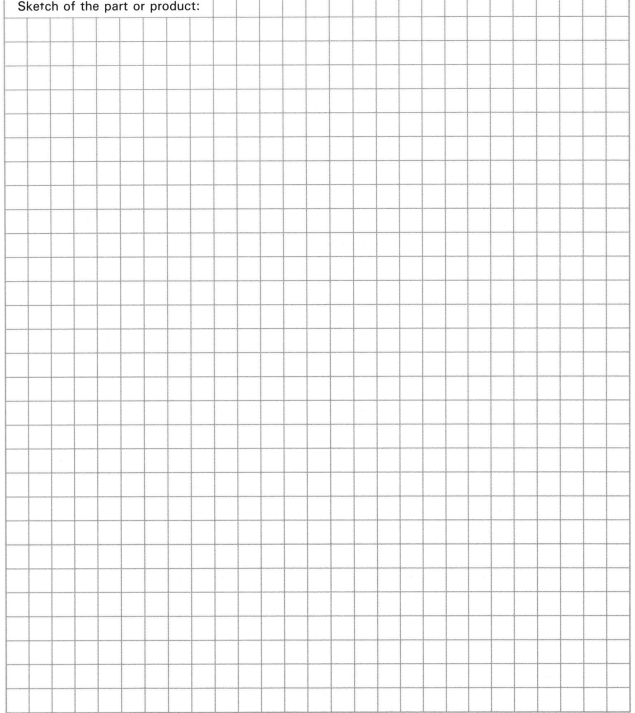

Sketch of the part or product:

*(Continued)*

# Activity 3-4, Chapter 9 (Continued)

Name _____

Separating process to be used: _____

As your teacher demonstrates the process, list the procedures you are to use and record the safety precautions that you are to follow.

| Procedure | Safety precaution |
|---|---|
| 1. | |
| 2. | |
| 3. | |
| 4. | |
| 5. | |
| 6. | |
| 7. | |
| 8. | |
| 9. | |
| 10. | |
| 11. | |
| 12. | |
| 13. | |
| 14. | |
| 15. | |
| 16. | |
| 17. | |
| 18. | |
| 19. | |
| 20. | |

*(Continued)*

**Activity 3-4, Chapter 9**
**(Continued)**

Name _____

Select one of the separating processes you used. Mark the appropriate box of each major step in the process. Then write a brief description that explains each step you used in completing the process.

Process to be described: _____

---

1. Cutting element used:  ☐ Single point    ☐ Multiple point    ☐ Other

| Describe the cutting element here: | Sketch of the cutting element: |
|---|---|
|  |  |

---

2. Method of producing motion:

Feed motion:      ☐ Linear    ☐ Rotary    ☐ Reciprocating

Cutting motion:    ☐ Linear    ☐ Rotary    ☐ Reciprocating

---

3. Method of holding and moving:

| Description of the method: | Sketch of the method: |
|---|---|
| a.  The tool: |  |
| b.  The work: |  |

---

*(Continued)*

Name _____

Select three separating processes you and your classmates completed. List their proper names in the boxes across the top of the chart. For each process, write a brief description about the five basic concepts listed on the left of the chart.

| | | | |
|---|---|---|---|
| Cutting element | | | |
| Types of motion — Cutting motion | | | |
| Types of motion — Feed motion | | | |
| Method of holding — Tool | | | |
| Method of holding — Work | | | |

## Activity 3-5: Conditioning Processes

## Chapter 10

Name _____

Date _____

Score _____

Period _____

You and your teacher will select a part or product to be manufactured in class. On the grid below, sketch the part or product. Then, on the following pages, complete the related charts.

Name of the part or product to be manufactured: _____

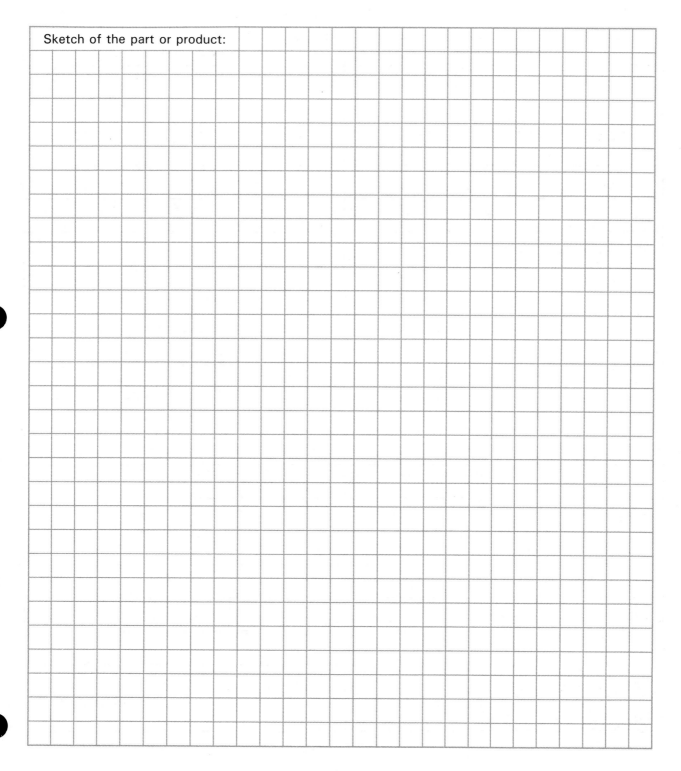

Sketch of the part or product:

*(Continued)*

**Activity 3-5, Chapter 10
(Continued)**

Name _____

Conditioning process to be used:_____

As your teacher demonstrates the process, list the procedures you are to use and record the safety precautions that you are to follow.

| Procedure | Safety precaution |
|---|---|
| 1. | |
| 2. | |
| 3. | |
| 4. | |
| 5. | |
| 6. | |
| 7. | |
| 8. | |
| 9. | |
| 10. | |
| 11. | |
| 12. | |
| 13. | |
| 14. | |
| 15. | |
| 16. | |
| 17. | |
| 18. | |
| 19. | |
| 20. | |

*(Continued)*

Name _____

Mark the appropriate box of each major step in the process. Then write a brief description that explains each step you used in completing the process. Finally, carefully observe and test the part after it is conditioned. Then record your observations.

Process to be described: _____

| | |
|---|---|
| 1. Type of process:   ☐ Thermal      ☐ Mechanical      ☐ Chemical | |

2. Type of property to be changed:

    ☐ Physical          ☐ Thermal          ☐ Acoustical

    ☐ Mechanical     ☐ Electrical/magnetic     ☐ Optical

    ☐ Chemical

3. Procedure:

4. Characteristics of the part after conditioning (visual appearance, hardness, stiffness, etc):

*(Continued)*

Name _____

Select two conditioning processes you or your classmates completed. List their proper names in the boxes across the top of the chart. For each process, write a brief description about the three factors listed on the left of the chart.

| | | |
|---|---|---|
| Property to be changed | | |
| Method of changing the property | ☐ Thermal<br>☐ Mechanical<br>☐ Chemical | ☐ Thermal<br>☐ Mechanical<br>☐ Chemical |
| Procedure followed | | |

## Activity 3-6: Assembling Processes

Name _____

Date _____

Score _____

Period _____

You and your teacher will select a part or product to be manufactured in class. In the space below, sketch how the parts are to be assembled. Then, on the following pages, complete the related charts.

Name the part or product to be manufactured: _____

*(Continued)*

Name _____

Method of assembly to be used: _____

As your teacher demonstrates the process, list the procedures you are to use and record the safety precautions that you are to follow.

| Procedure | Safety precaution |
|---|---|
| 1. | |
| 2. | |
| 3. | |
| 4. | |
| 5. | |
| 6. | |
| 7. | |
| 8. | |
| 9. | |
| 10. | |
| 11. | |
| 12. | |
| 13. | |
| 14. | |
| 15. | |
| 16. | |
| 17. | |
| 18. | |
| 19. | |
| 20. | |

*(Continued)*

# Activity 3-6, Chapter 11 (Continued)

Name _____

Mark the appropriate box for each major factor in the process. Then make a sketch or write a brief description as indicated in each part of the form.

Process to be described: _____

| | |
|---|---|
| 1. Type of process:     ☐ Bonding        ☐ Mechanical fastening | |
| 2. Agent used:<br><br>   ☐ Fastener<br><br>   ☐ Mechanical force<br><br>   ☐ Bonding agent | Sketch the agent here showing its relationship with the original parts to be assembled. |
| 3. Method of applying agent: | |
| 4. Joint used:<br><br>   ☐ T-joint<br><br>   ☐ Corner joint<br><br>   ☐ Angle joint<br><br>   ☐ Butt joint<br><br>   ☐ Skarf joint | Sketch the joint used here (Label the fastener or bonding area.): |
| Why is this joint appropriate for this application? | |

*(Continued)*

Name _____

Select three assembling processes you or your classmates completed. List their proper names in the boxes across the top of the chart. For each process, write a brief description about the basic factors listed on the left of the chart.

| | | | |
|---|---|---|---|
| Type of assembling process | ☐ Bonding<br>☐ Mechanical fastening | ☐ Bonding<br>☐ Mechanical fastening | ☐ Bonding<br>☐ Mechanical fastening |
| **Bonding** — Bonding agent | | | |
| **Bonding** — Method of bonding (List a brief procedure.) | | | |
| **Mechanical fastening** — Joint (Make a sketch.) | | | |
| **Mechanical fastening** — Fastening agent (List name and a brief procedure.) | | | |

## Activity 3-7: Finishing Processes          ## Chapter 12

Name _____          Score _____

Date _____          Period _____

You and your teacher will select a part or product to be manufactured in class. On the grid below, sketch the part or product. Then, on the following pages, complete the related charts.

Name of the part or product to be manufactured: _____

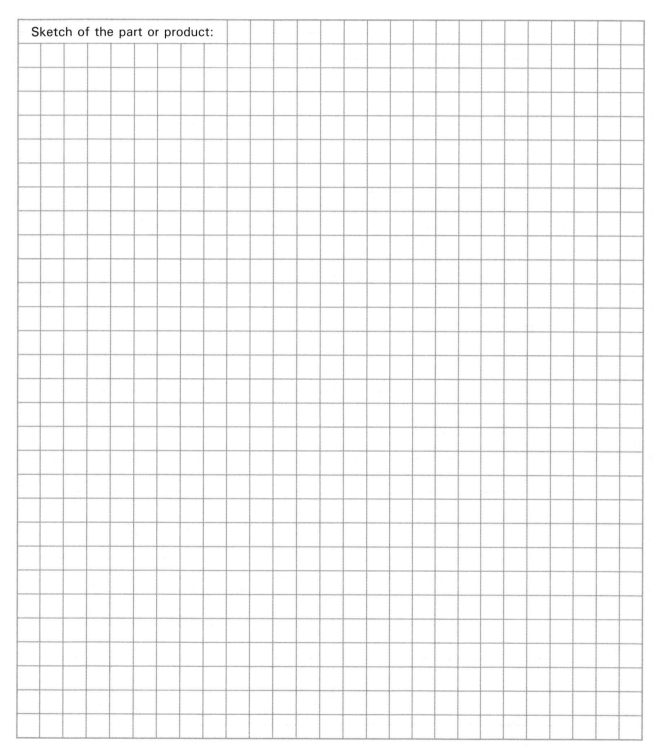

Sketch of the part or product:

*(Continued)*

# Activity 3-7, Chapter 12 (Continued)

Name _____

Finishing process to be used: _____

As your teacher demonstrates the process, list the procedures you are to use and record the safety precautions that you are to follow.

| Procedure | Safety precaution |
|---|---|
| 1. | |
| 2. | |
| 3. | |
| 4. | |
| 5. | |
| 6. | |
| 7. | |
| 8. | |
| 9. | |
| 10. | |
| 11. | |
| 12. | |
| 13. | |
| 14. | |
| 15. | |
| 16. | |
| 17. | |
| 18. | |
| 19. | |
| 20. | |

*(Continued)*

# Activity 3-7, Chapter 12
## (Continued)

Name _____

Mark the appropriate box for each major factor in the process. Then make a sketch or write a brief description as indicated in each part of the form.

Process to be described: _____

---

1. Type of process:  ☐ Coating  ☐ Surface conversion

---

2. Type of finishing material used:  ☐ Organic  ☐ Inorganic

   Name of the specific finishing material: _____

---

3. Method for preparing the material for finishing:

   ☐ Mechanical  ☐ Chemical

   Procedure (List the main steps below.):

---

4. Method for applying the finish:

   ☐ Brushing  ☐ Dipping  ☐ Plating

   ☐ Rolling  ☐ Spraying  ☐ Chemical deposition/conversion

   Procedure (List the main steps below.):

---

5. Why was this finish chosen?

---

*(Continued)*

Name _____

Select two finishing processes you or your classmates completed. List their proper names in the boxes across the top of the chart. For each process, write a brief description about the three factors listed on the left of the chart.

| | | |
|---|---|---|
| Finishing material | ☐ Inorganic    ☐ Organic<br><br>Name: | ☐ Inorganic    ☐ Organic<br><br>Name: |
| Material preparation (Check method used and give a brief procedure.) | ☐ Mechanical    ☐ Chemical | ☐ Mechanical    ☐ Chemical |
| Finish material application (Check method used and give a brief procedure.) | ☐ Brushing  ☐ Rolling  ☐ Spraying<br>☐ Dipping<br>☐ Other (specify technique): | ☐ Brushing  ☐ Rolling  ☐ Spraying<br>☐ Dipping<br>☐ Other (specify technique): |

# Section 4
# INTRODUCTION TO MANAGEMENT

**Activity 4-1: Management Technology**

Name _____

Date _____

**Chapter 13**

Score _____

Period _____

| ECONOMIC OPPORTUNITY |
|---|
| An enterprising person has noticed that greeting cards are a popular product. This person has determined that the greeting card business reaches a number of market segments. It has been decided that a study of the market is necessary to determine the demand for the enterprise and, also, to reduce the risks involved in entering the market. |

Form a group of five students and brainstorm answers to the following topics:

1.  Types and sizes of cards:

    a.  List major events for which greeting cards are sent:

    Holidays:

    _____

    _____

    _____

    _____

    _____

    _____

    _____

    _____

    _____

    Special events:

    _____

    _____

    _____

    _____

    _____

    _____

    _____

    _____

    _____

    b.  Sizes of common cards:

    Standard:            _____

    Studio (humorous):   _____

    Oversized:           _____

*(Continued)*

2. Card design:

   a. Your teacher will give you two greeting cards. Analyze them in terms of the following:

     • For what event was the card designed?

      Card A: _____

      Card B: _____

     • For whom is the card appropriate (age, gender, etc.)?

      Card A: _____

      Card B: _____

     • Why do you think the designer selected the colors and layout that were used?

      Card A: _____

      _____

      _____

      _____

      Card B: _____

      _____

      _____

      _____

     • How would you change the design to improve it?

      Card A: _____

      _____

      Card B: _____

      _____

      _____

*(Continued)*

b. Describe the approach (type of message, type of picture or art, use of humor, etc.) you would use for each of the events and/or people listed below.

| Person and event | Approach |
|---|---|
| Birthday (Five-year-old child) | |
| Birthday (Teenager) | |
| Birthday (Young adult) | |
| Birthday (Older adult) | |
| Friendship (Boyfriend or girlfriend) | |
| Graduation (High school graduate) | |
| Sympathy (Close friend) | |

*(Continued)*

| Person and event | Approach |
|---|---|
| Get well (Teenage friend) | |
| Get well (Grandparent) | |
| Mother's Day | |
| Father's Day | |
| Holiday: _____ (Youth) | |
| Holiday: _____ (Adult) | |
| Thank you (For a gift) | |

## Activity 4-2: Organization and Structure

**Chapter 14**

Name _____

Date _____

Score _____

Period _____

| BUSINESS OPPORTUNITY |
|---|
| You have decided to enter the greeting card market. Your company will focus on holiday and special event cards for the youth market. This market is defined as people from ages 15 through 18. You must decide on the type of ownership your company will have and the managerial structure that will best fit the company. |

Join a group of students in your class to organize and structure your company by completing the two tasks described below.

### Type of ownership

List the advantages and disadvantages for each of the three types of ownership listed below. Then circle the one you would recommend for this business.

| | Advantages | Disadvantages |
|---|---|---|
| Proprietorship | | |
| Partnership | | |
| Corporation | | |

**Activity 4-2, Chapter 14
(Continued)**

Name _____

## Managerial structure

List the major jobs involved in designing, producing, and selling your greeting cards.

1. _____     5. _____

2. _____     6. _____

3. _____     7. _____

4. _____     8. _____

Develop a company organization chart for your greeting card company that shows the "chain of command" from the company president to the workers. Note: You may wish to refer to the organization charts shown in Chapter 14 of the text as examples.

**Activity 4-3: Product-Centered Activities**    **Chapter 15**

Name _____    Score _____

Date _____    Period _____

| THE CHALLENGE |
|---|
| You have been employed as a designer for the Polar Scenes Greeting Card Company. Your first assignment is to prepare a card of your own choosing to show the Chief Designer what you can do. Use this form as you complete the task. |

**Audience and occasion:** For whom and what occasion is the card designed?

Gender:  ☐ Male        ☐ Female        ☐ Both

Age:  ☐ 6-10   ☐ 11-14   ☐ 15-18   ☐ 19-25   ☐ 26-35   ☐ 36-50   ☐ Over 50

Occasion:    ☐ Holiday (specify):_____

☐ Special event (specify): _____

**Theme:** What is the underlying message?

_____

_____

_____

**Approach:** What type of message and graphics will the card depict?

☐ Humor      ☐ Sincerity      ☐ Other (specify) _____

☐ Flowers/Trees      ☐ Wildlife      ☐ People      ☐ Automobiles      ☐ Buildings

☐ Other (specify) _____

**Message:** What will the card say?

_____

_____

_____

_____

_____

*(Continued)*

Name _____

**CARD DESIGN:** Prepare the card design, with graphics and type, on the layout form below.

Inside message (top of panel)

Front (Top of panel)

## Activity 4-4: Managed Support Areas

## Chapter 16

Name _____

Date _____

Score _____

Period _____

| THE TASK |
|---|
| The Polar Scene Greeting Card Company has chosen your greeting card for their new line. The Chief Designer has asked you to prepare a manufacturing procedure (operation sheet) so that material and labor costs can be estimated. |

List the steps it would take to manufacture the greeting card. Start with a blank sheet of paper and end with 12 cards and envelopes placed in a box that will be displayed in a card shop.

| Step number | Operation | Number of workers |
|---|---|---|
|  |  |  |
|  |  |  |
|  |  |  |
|  |  |  |
|  |  |  |
|  |  |  |
|  |  |  |
|  |  |  |
|  |  |  |
|  |  |  |
|  |  |  |
|  |  |  |
|  |  |  |
|  |  |  |
|  |  |  |
|  |  |  |
|  |  |  |
|  |  |  |
|  |  |  |
|  |  |  |

*(Continued)*

**Activity 4-4, Chapter 16 (Continued)**

Name _____

Compare your list with those of others in your class.

1. In what ways are the lists alike?

_____

_____

_____

_____

_____

_____

2. In what ways are the lists different?

_____

_____

_____

_____

_____

_____

3. How could you revise your list to make the operation more efficient?

_____

_____

_____

_____

_____

_____

_____

_____

_____

_____

_____

# Section 5
# MANUFACTURING ENTERPRISE

## Activity 5-1: Articles of Incorporation          Chapter 17

Name _____          Score _____

Date _____          Period _____

Your class has decided to form a manufacturing corporation. Complete the articles of incorporation below.

| ARTICLES OF INCORPORATION |
|---|
| State (or Province) of _____ |
| **ARTICLE 1 — Name** |
| The name of the Corporation is_____ |
| **ARTICLE 2 — Purposes** |
| The purposes for which the Corporation is formed are:<br><br>(1) _____<br><br>(2) _____<br><br>(3) _____ |
| **ARTICLE 3 — Period of Existence** |
| The period of time during which the Corporation will continue operations is:<br><br>_____ |
| **ARTICLE 4 — Principal Office** |
| The Post Office address of the principal office of the Corporation is:<br><br>_____<br>(Number and Street or Building)<br><br>_____<br>(City, State and Zip Code) |
| **ARTICLE 5 — Shares** |
| Total number of shares the Corporation has authority to issue is _____.<br><br>The number of shares which the Corporation designates as having par value is _____ with a par value of $_____ per share.<br><br>The number of shares which the Corporation designates as without par value _____. |

*(Continued)*

Name _____

| ARTICLE 6 — Directors |
|---|

**Section 1: Number of Directors:**

The initial Board of Directors is composed of _____ members. The number of directors may be from time to time fixed by the By-Laws of the Corporation at any number.

**Section 2: Names and Post Office Addresses of the Directors:**

The names and post office addresses of the initial Board of Directors of the Corporation are:

| NAME | NUMBER AND STREET OR BUILDING | CITY | STATE | ZIP CODE |
|---|---|---|---|---|
| | | | | |
| | | | | |
| | | | | |
| | | | | |
| | | | | |

| ARTICLE 7 — Officers |
|---|

The names and post office addresses of the incorporators of the Corporation are:

| NAME | NUMBER AND STREET OR BUILDING | CITY | STATE | ZIP CODE |
|---|---|---|---|---|
| | | | | |
| | | | | |
| | | | | |
| | | | | |

IN WITNESS WHEREOF, the undersigned, being the incorporators designated in Article 7, execute these articles of Incorporation and certify to the truth of the facts herein stated, the _____ day of _____, 19_____.

_____
(Written Signature)

_____
(Written Signature)

_____
(Printed Signature)

_____
(Printed Signature)

## Activity 5-2: Issuing Stock Certificates

**Chapter 17**

Name _____

Date _____

Score _____

Period _____

Duplicate the stock certificate below. Then complete the stock certificates and issue them to stockholders.

Certificate No: _____

No. of shares _____

Issue price of each

share: _____

**STOCK REGISTER**

(Separate from Certificate

when sold.)

Stockholder's address at

redemption date:

_____
(Name)

_____
(Street)

_____
(City, State)

_____
(Zip Code)

Date: _____

Number of Non-Legal Shares ☐

Certificate Number _____

Redemption Date: _____

Number of Shares - Punch Out

| 0 | 0 |
|---|---|
| 1 | 1 |
| 2 | 2 |
| 3 | 3 |
| 4 | 4 |
| 5 | 5 |
| 6 | 6 |
| 7 | 7 |
| 8 | 8 |
| 9 | 9 |

is a simulated, non-legal corporation of the manufacturing

class of _____
(Name)

school in _____
(City, State or Province)

This certificate is witness that _____
is a legal owner of _____ shares of no par value stock
in the above named company. Stock will be redeemed at
book value on the redemption date above and is void if not
presented to the transfer agent within seven (7) days after
that date.

_____
(President)

_____
(Vice-President, Financial Affairs)

# Activity 5-3: Stockholder's Ledger

## Chapter 17

Name _____

Score _____

Date _____

Period _____

Complete the following ledger of stocks that are issued for your enterprise.

Company name: _____

Recorder: _____

| Date | Certificate number | Stockholder's name | Address | Number of shares |
|------|------|------|------|------|
|  |  |  |  |  |
|  |  |  |  |  |
|  |  |  |  |  |
|  |  |  |  |  |
|  |  |  |  |  |
|  |  |  |  |  |
|  |  |  |  |  |
|  |  |  |  |  |
|  |  |  |  |  |
|  |  |  |  |  |
|  |  |  |  |  |
|  |  |  |  |  |
|  |  |  |  |  |
|  |  |  |  |  |
|  |  |  |  |  |
|  |  |  |  |  |
|  |  |  |  |  |
|  |  |  |  |  |
|  |  |  |  |  |
|  |  |  |  |  |
|  |  |  |  |  |
|  |  |  |  |  |
|  |  |  |  |  |
|  |  |  |  |  |
|  |  |  |  |  |

# Activity 5-4: Budget Request

## Chapter 17

Name _____

Date _____

Score _____

Period _____

Using the form below, complete the following budget request.

Company name: _____

Prepared by: _____

| DEPARTMENT |
|---|
| Department name: _____ |
| Manager in charge: (Name) _____ (Title) _____ |

| MATERIALS | | |
|---|---|---|
| Item | Quantity | Estimated cost |
| | | |
| | | |
| | | |
| | | |
| | | |
| TOTAL MATERIAL COST | | |

| LABOR COSTS | | | |
|---|---|---|---|
| Type | Labor Rate | Hours/Days | Total |
| Managerial/Administration | | | |
| Design/Engineering | | | |
| Production | | | |
| Sales | | | |
| TOTAL LABOR COST | | | |

| OVERHEAD | | |
|---|---|---|
| Estimate using the following formula: Overhead = material cost + labor cost × 125% | Material cost:_____ Labor cost:_____ | OVERHEAD _____ |

# Activity 5-5: Master Budget

# Chapter 17

Name _____

Score _____

Date _____

Period _____

Using the form below, prepare a master budget for your company.

Company name: _____

Prepared by: _____

| BUDGET PERIOD | |
|---|---|
| Starting date: | Ending date: |

| BUDGET | |
|---|---|

**SALES INCOME** $_____

**EXPENSES**

  **Production Expenses**

    Materials    $_____

    Labor    $_____

    Overhead    $_____

    **Total Production Expenses**    $_____

  **Marketing Expenses**

    Advertising and Promotion    $_____

    Selling    $_____

    Other    $_____

    **Total Marketing Expenses**    $_____

  **General Expenses**

    Administration    $_____

    Other    $_____

    **Total General Expense**    $_____

**TOTAL EXPENSES**    $_____

**PROFIT OR (LOSS) BEFORE TAXES**    $_____

# Activity 5-6: Product Profile

## Chapter 18

Name _____

Date _____

Score _____

Period _____

---

### THE TASK

Develop a profile for products that are appropriate for your company by completing all sections of this form.

---

### PRODUCT CONSIDERATIONS

Type of product: _____

Need for the product: _____

How is it used (utility)? _____

Expected life: _____

---

### MARKET CONSIDERATIONS

Who is the market (Who will buy the product?)?_____

How big is the market? _____

How much competition is there? _____

---

### PRODUCTION CONSIDERATIONS

How complex is the product? _____

Material needed:_____

Equipment available: _____

_____

Tooling (maximum cost and time to fabricate): _____

---

### FINANCIAL CONSIDERATIONS

Material cost (maximum): _____

Selling price (desired range): _____

Profit margin (desired range in percentage of selling price): _____

Name _____    Score _____

Date _____    Period _____

Product name: _____

| THE TASK |
|---|
| Photocopy this page so that you have at least five of the isometric grid below. Then develop at least five rough sketches of a product that will fit your company's product profile. Each sketch should be one solution to the product needs. |

## Activity 5-8: Refined Sketches          ## Chapter 18

Name _____          Score _____

Date _____          Period _____

Product name: _____

| THE TASK |
|---|
| Select your best rough sketch. Refine it by adding detail, improving its shape and proportions, smoothing styling lines, etc. |

# Activity 5-9: Planning Board Evaluation Checklist   Chapter 18

Name _____

Score _____

Date _____

Period _____

| THE TASK |
|---|
| Rate products presented by members of the class to determine which ones have the best chance for earning a profit for the company. Rate each product on the four categories included in the chart below. Complete the product profile using a scale of 1 to 5, with 5 being the best rating in each category. |

| Product number | Description | RATINGS | | | | Total |
|---|---|---|---|---|---|---|
| | | Design | Market | Production | Financial | |
| | | | | | | |
| | | | | | | |
| | | | | | | |
| | | | | | | |
| | | | | | | |
| | | | | | | |
| | | | | | | |
| | | | | | | |
| | | | | | | |
| | | | | | | |
| | | | | | | |
| | | | | | | |
| | | | | | | |
| | | | | | | |
| | | | | | | |
| | | | | | | |
| | | | | | | |
| | | | | | | |
| | | | | | | |
| | | | | | | |

## Activity 5-10: Second Planning Board Assignments

**Chapter 18**

Name _____

Date _____

Score _____

Period _____

Complete the second planning board assignments form below.

Product: _____     Product development group number: _____

| Name | Responsibility | Due date | Notes |
|------|---------------|----------|-------|
| | Group leader | | |
| | Data summary sheet (Compiler) | | |
| | Market survey (Developer) | | |
| | Presentation chart #1 (Developer) | | |
| | Presentation chart #2 (Developer) | | |
| | Competitive study (Study manager) | | |
| | Drawing (name) (Assign drawings to each member) | | |
| 1 | | | |
| 2 | | | |
| 3 | | | |
| 4 | | | |
| 5 | | | |
| | Bill of materials (Developer) | | |
| | Cost estimate (Study manager) | | |
| | Other (Specify) | | |

# Activity 5-11: Product Drawings

**Name** _____

**Date** _____

## Chapter 18

**Score** _____

**Period** _____

In the space provided, prepare the final product drawing.

| Product name:<br>Part name: | (Company name) | Drafter:<br>Scale:      Date: |
|---|---|---|

Copyright Goodheart-Willcox Co., Inc.

# Activity 5-12: Bill of Materials

## Chapter 18

Name _____

Date _____

Score _____

Period _____

Complete the bill of materials form below.

Product: _____     Product development group number:_____

| Part number | Quantity needed | Part name | Size | | | Material |
|---|---|---|---|---|---|---|
| | | | Thickness | Width | Length | |
| | | | | | | |
| | | | | | | |
| | | | | | | |
| | | | | | | |
| | | | | | | |
| | | | | | | |
| | | | | | | |
| | | | | | | |
| | | | | | | |
| | | | | | | |
| | | | | | | |
| | | | | | | |
| | | | | | | |
| | | | | | | |
| | | | | | | |
| | | | | | | |
| | | | | | | |
| | | | | | | |
| | | | | | | |
| | | | | | | |

## Activity 5-13: Material Cost Estimate

**Chapter 18**

Name _____

Score _____

Date _____

Period _____

Complete the material cost estimate below.

Product: _____     Compiler: _____

| Name of the part | Quantity needed | Vendor name | Vendor quote | Best price |
|---|---|---|---|---|
|  |  |  |  |  |
|  |  |  |  |  |
|  |  |  |  |  |
|  |  |  |  |  |
|  |  |  |  |  |
|  |  |  |  |  |
|  |  |  |  |  |
|  |  |  |  |  |

## Activity 5-14: Flow Process Chart

Name _____

Date _____

Score _____

Period _____

Complete the flow process chart below.

| Product name: _____ | | Prepared by: _____ |
| --- | --- | --- |
| Flow begins: _____ | Flow ends: _____ | Approved by: _____ |
| Process symbols and number of times used | ◯ Operation _____  ⇨ Transportation _____ | ▢ Inspection _____   ▽ Storage _____ |  D Delay _____ |

| Task no. | Process symbol | Description of task | Machine required | Tooling required |
| --- | --- | --- | --- | --- |
|  | ◯⇨▢D▽ |  |  |  |
|  | ◯⇨▢D▽ |  |  |  |
|  | ◯⇨▢D▽ |  |  |  |
|  | ◯⇨▢D▽ |  |  |  |
|  | ◯⇨▢D▽ |  |  |  |
|  | ◯⇨▢D▽ |  |  |  |
|  | ◯⇨▢D▽ |  |  |  |
|  | ◯⇨▢D▽ |  |  |  |
|  | ◯⇨▢D▽ |  |  |  |
|  | ◯⇨▢D▽ |  |  |  |
|  | ◯⇨▢D▽ |  |  |  |
|  | ◯⇨▢D▽ |  |  |  |
|  | ◯⇨▢D▽ |  |  |  |
|  | ◯⇨▢D▽ |  |  |  |
|  | ◯⇨▢D▽ |  |  |  |
|  | ◯⇨▢D▽ |  |  |  |
|  | ◯⇨▢D▽ |  |  |  |
|  | ◯⇨▢D▽ |  |  |  |
|  | ◯⇨▢D▽ |  |  |  |

## Activity 5-15: Plant Layout

**Chapter 19**

Name _____

Date _____

Score _____

Period _____

Product name: _____

Prepared by: _____

Draw the outline of the laboratory on the grid below. Photocopy the symbols on the next page. Cut them out and arrange them within the boundaries of the laboratory to develop an efficient plant layout.

*(Continued)*

| OPERATORS | MACHINES | BENCHES AND CONVEYORS |
|---|---|---|

SCROLL SAW

DRILL PRESS

BAND SAW

WORK BENCH

4-STATION WORK BENCH

SCROLL SAW

DRILL PRESS

BAND SAW

WORK BENCH

4-STATION WORK BENCH

DISC AND BELT SANDER

BELT SANDER

DISC SANDER

WORKBENCH

4-STATION WORK BENCH

GRINDER

TABLE SAW

JOINTER

LATHE

LATHE

WORK BENCH

4-STATION WORK BENCH

SPOT WELDER

SQUARING SHEAR

ROD BENDER

BOX AND PAN BRAKE

CONVEYOR

CONVEYOR

CONVEYOR

CONVEYOR

EXTRA MACHINES AND BENCHES (WRITE IN NAMES)

SMALL TABLE

SMALL TABLE

SMALL TABLE

SMALL TABLE

SMALL TABLE

SMALL TABLE

# Activity 5-16: Tooling Design

**Chapter 19**

Name _____

Date _____

Score _____

Period _____

Tooling designer: _____

Product name: _____

| THE TASK |
|---|

This tooling will be used to (task)_____

    on the (part name or names) _____.

Type of tooling:    ☐ Jig    ☐ Fixture    ☐ Pattern    ☐ Template

      for:    ☐ Fabrication    ☐ Assembly

Machine tooling will fit: _____

  Note: Draw the tooling in black, showing the tool (cutter, drill, etc.) position in red, and the work in green.

## Activity 5-17: Inspection Gage Design　　Chapter 19

Name _____

Date _____

Score _____

Period _____

Gage designer: _____

Product name: _____

| THE TASK |
| --- |

This gage will inspect (task or feature) _____

on the (part name or names) _____.

Note: Draw the gage in black and the work in green.

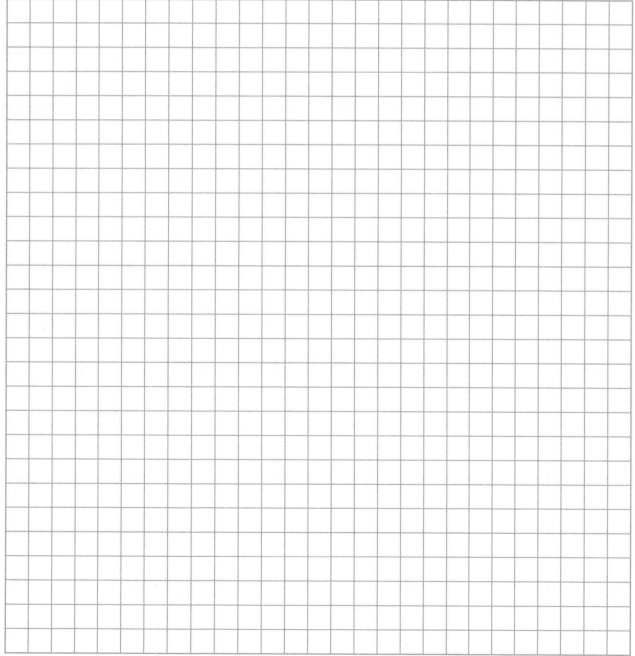

# Activity 5-18: Quality Control Inspection Instructions

## Chapter 19

Name _____

Score _____

Date _____

Period _____

Complete the quality control inspection instruction form below.

Product: _____ Part: _____ Developed by: _____

| Operation Number | Characteristic to inspect | Method of inspecting | Gage number |
|---|---|---|---|
| | | | |
| | | | |
| | | | |
| | | | |
| | | | |
| | | | |
| | | | |
| | | | |
| | | | |
| | | | |
| | | | |

## Activity 5-19: Employee Requisition

Name _____

Date _____

Score _____

Period _____

Complete the employee requisition form below.

---

**EMPLOYEE REQUISITION**

Department head: Complete this form for each employee requested and return it to the personnel office.

JOB TITLE: -----------------------------------------------------

BRIEF DESCRIPTION OF DUTIES: -------------------------------------

-----------------------------------------------------------------

-----------------------------------------------------------------

-----------------------------------------------------------------

-----------------------------------------------------------------

-----------------------------------------------------------------

EXPERIENCE REQUIRED: ---------------------------------------------

-----------------------------------------------------------------

-----------------------------------------------------------------

-----------------------------------------------------------------

TO WHOM WILL THE EMPLOYEE REPORT?---------------------------------

DATE EMPLOYEE IS NEEDED:------------------------------------------

PREPARED BY: -------------------------- DEPARTMENT: --------------

---

# Activity 5-20: Job Description

**Chapter 20**

Name _____

Score _____

Date _____

Period _____

Complete the job description below.

| | |
|---|---|
| Job title: ─────────────────────────────────── | Job code: ──────── |
| Prepared by: ───────────────────────────────── | Date: ──────── |

**Principal duties:**




**Materials or product worked on:**




**Typical equipment used:**




**Typical tools and measuring instruments used:**

# Activity 5-21: Application for Employment      Chapter 20

Name _____      Score _____

Date _____      Period _____

Complete the employment application below.

---

## PERSONAL INFORMATION

Name _____    Social Security Number _____
    (Last)      (First)      (Middle Initial)

Address_____    Telephone number _____

_____

How long have you lived at the above address? _____

Position desired: _____    Date you are available: _____

## EMPLOYMENT RECORD

List all employers with most recent first. Include summer and part-time positions.

| Employer | Dates employed From | To | Job title | Reason for leaving |
|---|---|---|---|---|
| | | | | |
| | | | | |
| | | | | |
| | | | | |

## EDUCATION RECORD

List all schools you attended and the level of education completed for each.

| Educational institution | Dates From | To | Completed Yes | No |
|---|---|---|---|---|
| (Grade School) | | | | |
| (Middle/Junior High) | | | | |
| (High School) | | | | |

## REFERENCES

| Name | Address | Occupation |
|---|---|---|
| | | |
| | | |

**Activity 5-22: The Job Interview**          **Chapter 20**

Name _____     Score _____

Date _____     Period _____

Questions that are often asked during a job interview are listed below. Answer these questions as you would during an interview. (These questions can also be used as you interview employees for your company.)

Job for which you are applying: _____

Company: _____

1. Why do you want to work for this company? _____

   _____

2. Do you think you will like this kind of work? _____ Why? _____

   _____

   _____

3. How would you describe yourself? _____

   _____

   _____

   _____

4. What are your best subjects in school? _____

   _____

5. What are your worst subjects in school? _____

   _____

6. What other jobs have you had? _____

   _____

7. Have you ever been fired from a job? If so, why? _____

   _____

8. What is your best qualification for this job? _____

   _____

   _____

   _____

9. What are your future plans? _____

   _____

10. Why should I hire you? _____

    _____

    _____

    _____

    _____

    _____

## Activity 5-23: Employment Examination

Name _____

Date _____

### Chapter 20

Score _____

Period _____

Complete the sample employment examination below.

This is a test of your ability to do common calculations and read drawings commonly used in the production area of this company.

| | ANSWERS |
|---|---|

1. How many degrees are there in a circle?

1. _____

2. How many degrees are there in a right triangle?

2. _____

3. If a circle has a radius of 5/32″, what is its diameter?

3. _____

4. What is the decimal equivalent of:

| DECIMAL EQUIVALENTS | | | |
|---|---|---|---|
| 1/32 | 0.01563 | 9/32 | 0.28125 |
| 1/16 | 0.06250 | 5/16 | 0.31250 |
| 3/32 | 0.09375 | 11/32 | 0.34375 |
| 1/8 | 0.12500 | 3/8 | 0.37500 |
| 5/32 | 0.15625 | 13/32 | 0.40625 |
| 3/16 | 0.18750 | 7/17 | 0.43750 |
| 7/32 | 0.21875 | 15/32 | 0.46875 |
| 1/4 | 0.25000 | 1/2 | 0.50000 |

a. 1/16″

4a. _____

b. 5/32″

4b. _____

c. 15/32″

4c. _____

5. This is a simple blueprint. Determine the dimensions for each letter indicated.

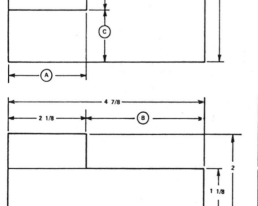

5 A _____

5B _____

5C _____

5D _____

5E _____

5F _____

5G _____

# Activity 5-24: Safety Poster Design Sheet    Chapter 20

Name _____    Score _____

Date _____    Period _____

To promote safety within your company, design a safety poster.

Safety campaign theme: _____

Safety slogan:_____

Poster prepared by: _____

NOTE: Indicate type sizes and styles, colors, and the type and location of all graphics.

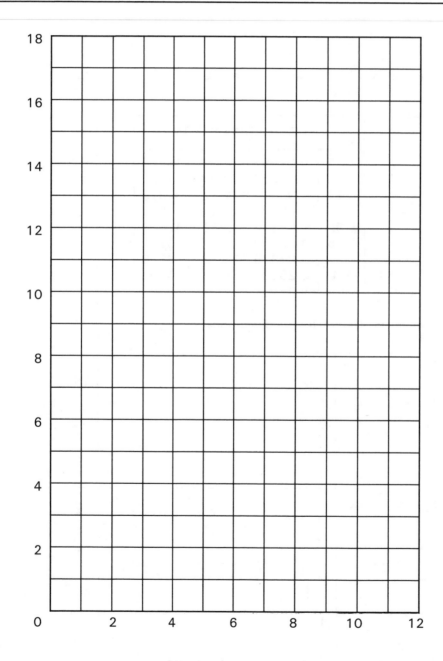

## Activity 5-25: Purchase Requisition                     Chapter 20

Name _____          Score _____

Date _____          Period _____

Complete the form below for each item needed and forward it to the purchasing office for action.

| ORDERING INFORMATION | | | |
|---|---|---|---|
| (Recommended vendors and addresses) | | | |

| ITEMS NEEDED | | | |
|---|---|---|---|
| Quantity | Catalog number | Description | Estimated cost |
|  |  |  |  |
|  |  |  |  |
|  |  |  |  |
|  |  |  |  |
|  |  |  |  |
|  |  |  |  |
|  |  |  |  |
|  |  |  |  |
|  |  |  |  |
|  |  |  |  |
|  |  |  |  |
|  |  |  |  |
|  |  |  |  |
|  |  |  |  |
|  |  |  |  |
|  |  |  |  |
|  |  |  |  |
|  |  |  |  |
|  |  |  |  |
|  |  |  |  |

# Activity 5-26: Purchase Order

**Chapter 20**

Name _____

Date _____

Score _____

Period _____

Complete the purchase order below.

| ORDERING INFORMATION | | | |
|---|---|---|---|
| (Vendor and address) | | (Ship to) | |

| ITEMS NEEDED | | | |
|---|---|---|---|
| Quantity | Catalog number | Description | Cost |
| | | | |
| | | | |
| | | | |
| | | | |
| | | | |
| | | | |
| | | | |
| | | | |
| | | | |
| | | | |
| | | | |
| | | | |
| | | | |
| | | | |
| | | | |
| | | | |
| | | | |
| | | | |
| | | | |
| | | | |

**Note:** One copy of this order is distributed as follows: Vendor, ordering department, purchasing office, and receiving dock.

Copyright Goodheart-Willcox Co., Inc.

## Activity 5-27:  Daily Production Schedule and Report

Name _____          Score _____

Date _____          Period _____

Complete the daily production schedule and report below.

## DAILY PRODUCTION SCHEDULE AND REPORT

Department: _____          Reported by: _____

| Part or product | Schedule | | Production | | | |
|---|---|---|---|---|---|---|
| | Part or assembly description | Number scheduled | Number produced | Scrap loss | Number of good products | Hours worked |
| | | | | | | |
| | | | | | | |
| | | | | | | |
| | | | | | | |
| | | | | | | |
| | | | | | | |
| | | | | | | |
| | | | | | | |
| | | | | | | |
| | | | | | | |
| | | | | | | |
| | | | | | | |
| | | | | | | |
| | | | | | | |
| | | | | | | |
| | | | | | | |
| | | | | | | |

## Activity 5-28: Individual Quality Goal

## Chapter 21

Name _____

Date _____

Score _____

Period _____

On the form below, develop a bar chart showing the daily percentage of parts you produced that passed inspection. Indicate your goal, and fill in boxes on the chart to the level of your achievement.

Operation: _____

Goal:_____

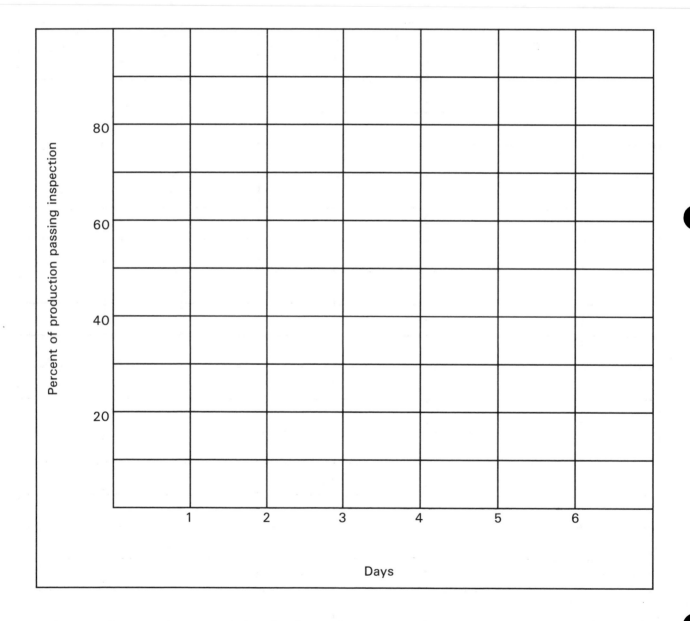

# Activity 5-29: Inspection Report

## Chapter 21

Name _____

Date _____

Score _____

Period _____

Complete the inspection report below.

## INSPECTION REPORT

Department: _____

Reported by: _____

| Part or product | | | Inspection results | | | | | | | |
|---|---|---|---|---|---|---|---|---|---|---|
| Part or product | Part or product name | Number produced | Number passed | Number rejected | | | | | | |
| | | | | (Reason) | (Reason) | (Reason) | (Reason) | (Reason) | (Reason) | |
| | | | | | | | | | | |
| | | | | | | | | | | |
| | | | | | | | | | | |
| | | | | | | | | | | |
| | | | | | | | | | | |
| | | | | | | | | | | |
| | | | | | | | | | | |
| | | | | | | | | | | |
| | | | | | | | | | | |
| | | | | | | | | | | |
| | | | | | | | | | | |
| | | | | | | | | | | |
| | | | | | | | | | | |
| | | | | | | | | | | |
| | | | | | | | | | | |
| | | | | | | | | | | |
| | | | | | | | | | | |
| | | | | | | | | | | |
| | | | | | | | | | | |

## Activity 5-30: Inspection Tags

Name _____

Date _____

**Chapter 21**

Score _____

Period _____

Duplicate these tags on colored index paper according to the following key.

White = Passed

# PASSED
### READY FOR:

☐ Further manufacture     ☐ Assembly

☐ Finishing     ☐ Packaging

☐ Shipping

SIGNED: _____

# REWORK

DIRECTIONS: _____

_____

_____

SIGNED: _____

Yellow = Rework

Pink = Reject

# REJECT

REASON: _____

_____

_____

SIGNED: _____

Copyright Goodheart-Willcox Co., Inc.

## Activity 5-31: Employee Time Cards

Name _____

Date _____

Score _____

Period _____

Duplicate the time card below on index paper. Use it to record your time as you participate in the manufacturing enterprise.

| TIME CARD | | | |
|---|---|---|---|
| NAME: _____ | | | |
| POSITION: _____ | | | |
| EMPLOYEE NUMBER: _____ | | | |
| WEEK — FROM: _____ TO: _____ | | | |
| DATE | IN | OUT | TIME |
| | | | |
| | | | |
| | | | |
| | | | |
| | | | |
| | | | |
| | | | |
| | | TOTAL: | |

# Activity 5-32: Payroll Sheet

## Chapter 21

Name _____

Date _____

Complete the payroll sheet below using data collected from time cards.

Company name:_____    Date: _____

For the week from: _____ to: _____    Prepared by: _____

| EMPLOYEE NAME | (1) REGULAR HOURS | (2) OVERTIME HOURS | (3) OVERTIME × 1.5 | TOTAL STD. HOURS (1) + (3) |
|---|---|---|---|---|
| | | | | |
| | | | | |
| | | | | |
| | | | | |
| | | | | |
| | | | | |
| | | | | |
| | | | | |
| | | | | |
| | | | | |
| | | | | |
| | | | | |
| | | | | |
| | | | | |
| | | | | |
| | | | | |
| | | | | |
| | | | | |
| | | | | |
| | | | | |
| | | | | |

## Activity 5-33: Union Organizing Leaflet Design    Chapter 22

Name _____    Score _____

Date _____    Period _____

Design a leaflet that could be used in promoting the organization of a union.

Theme: _____

Slogan: _____

Prepared by: _____

NOTE: Indicate types sizes and styles, colors, and the type and location of all graphics.

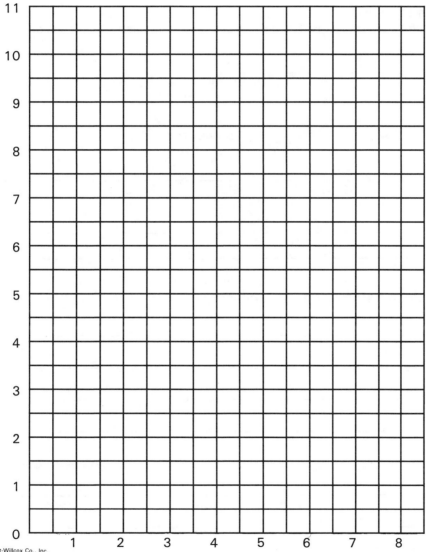

**Activity 5-34: Union Authorization Card**       **Chapter 22**

Name _____       Score _____

Date _____       Period _____

Complete the union authorization card below.

---

# AUTHORIZATION FOR REPRESENTATION BY

_____
(Name of the union)

I desire to be represented by the _____ union and hereby
designate said union as my bargaining agent in matters of wages, hours, and conditions of employment.

_____       _____
(Date)                         (Signature - do not print)

                       _____
                                     (Home address)

I am employed by:      _____
                         (City)       (State)      (Zip Code)       (Phone)

_____       _____       _____
(Name of company)              (Job title)                    (Department)

---

Explain why a card, such as this one, is necessary.

_____

_____

_____

_____

_____

_____

_____

_____

Name _____     Score _____

Date _____     Period _____

Complete the reproductionof an actual NLRB form below.

| Form NLRB-502 (11-64) | UNITED STATES OF AMERICA NATIONAL LABOR RELATIONS BOARD | Form Approved. Budget Bureau No. 64-R002.14 |
|---|---|---|
| | **PETITION** | **DO NOT WRITE IN THIS SPACE** CASE NO. |
| INSTRUCTIONS.—Submit an original and four (4) copies of this Petition to the NLRB Regional Office in the Region in which the employer concerned is located. If more space is required for any one item, attach additional sheets, numbering item accordingly. | | DATE FILED |

The Petitioner alleges that the following circumstances exist and requests that the National Labor Relations Board proceed under its proper authority pursuant to Section 9 of the National Labor Relations Act.

1. Purpose of this Petition *(If box RC, RM, or RD is checked and a charge under Section 8(b)(7) of the Act has been filed involving the Employer named herein, the statement following the description of the type of petition shall not be deemed made.)*

*(Check one)*

☐ **RC-CERTIFICATION OF REPRESENTATIVE** —A substantial number of employees wish to be represented for purposes of collective bargaining by Petitioner and Petitioner desires to be certified as representative of the employees.

☐ **RM-REPRESENTATION (EMPLOYER PETITION)**—One or more individuals or labor organizations have presented a claim to Petitioner to be recognized as the representative of employees of Petitioner.

☐ **RD-DECERTIFICATION** — A substantial number of employees assert that the certified or currently recognized bargaining representative is no longer their representative.

☐ **UD-WITHDRAWAL OF UNION SHOP AUTHORITY**—Thirty percent (30%) or more of employees in a bargaining unit covered by an agreement between their employer and a labor organization desire that such authority be rescinded.

☐ **UC-UNIT CLARIFICATION**—A labor organization is currently recognized by employer, but petitioner seeks clarification of placement of certain employees: *(Check one)* ☐ In unit not previously certified

☐ In unit previously certified in Case No. _____.

☐ **AC-AMENDMENT OF CERTIFICATION**—Petitioner seeks amendment of certification issued in Case No._____.

*Attach statement describing the specific amendment sought.*

| 2. NAME OF EMPLOYER | EMPLOYER REPRESENTATIVE TO CONTACT | PHONE NO. |
|---|---|---|

3. ADDRESS(ES) OF ESTABLISHMENT(S) INVOLVED *(Street and number, city, State, and ZIP Code)*

| 4a. TYPE OF ESTABLISHMENT *(Factory, mine, wholesaler, etc.)* | 4b. IDENTIFY PRINCIPAL PRODUCT OR SERVICE |
|---|---|

| 5. Unit Involved, *(In UC petition, describe PRESENT bargaining unit and attach description of proposed clarification.)* | 6a. NUMBER OF EMPLOYEES IN UNIT: |
|---|---|
| Included | PRESENT_____ |
| | PROPOSED (BY UC/AC) _____ |
| Excluded | 6b. IS THIS PETITION SUPPORTED BY 30% OR MORE OF THE EMPLOYEES IN THE UNIT?* ☐ YES ☐ NO *Not applicable in RM, UC, and AC |

*(If you have checked box RC in 1 above, check and complete EITHER item 7a or 7b, whichever is applicable)*

7a. ☐ Request for recognition as Bargaining Representative was made on _____and Employer *(Month, day, year)*

declined recognition on or about _____ *(If no reply received, so state)* *(Month, day, year)*

7b. ☐ Petitioner is currently recognized as Bargaining Representative and desires certification under the act.

*(Continued)*

---

8. Recognized or Certified Bargaining Agent *(If there is none, so state)*

| NAME | AFFILIATION |
|---|---|
| | |
| ADDRESS | DATE OF RECOGNITION OR CERTIFICATION |

| 9. DATE OF EXPIRATION OF CURRENT CONTRACT, IF ANY *(Show month, day, and year)* | 10. IF YOU HAVE CHECKED BOX UD IN 1 ABOVE, SHOW HERE THE DATE OF EXECUTION OF AGREEMENT GRANTING UNION SHOP *(Month, day, and year)* |
|---|---|

| 11a. IS THERE NOW A STRIKE OR PICKETING AT THE EMPLOYER'S ESTABLISH-MENT(S) INVOLVED?  YES _____ NO _____ | 11b. IF SO, APPROXIMATELY HOW MANY EMPLOYEES ARE PARTICIPATING? |
|---|---|

11c. THE EMPLOYER HAS BEEN PICKETED BY OR ON BEHALF OF ............................................................................., A LABOR
*(Insert name)*

ORGANIZATION, OF .................................................................. SINCE ................................
*(Insert address)*          *(Month, day, year)*

12. ORGANIZATIONS OR INDIVIDUALS OTHER THAN PETITIONER (AND OTHER THAN THOSE NAMED IN ITEMS 8 AND 11c), WHICH HAVE CLAIMED RECOGNITION AS REPRESENTATIVES AND OTHER ORGANIZATIONS AND INDIVIDUALS KNOWN TO HAVE A REPRESENTATIVE INTEREST IN ANY EMPLOYEES IN THE UNIT DESCRIBED IN ITEM 5 ABOVE. (IF NONE, SO STATE.)

| NAME | AFFILIATION | ADDRESS | DATE OF CLAIM *(Required only if Petition is filed by Employer)* |
|---|---|---|---|
| | | | |
| | | | |

I declare that I have read the above petition and that the statements therein are true to the best of my knowledge and belief.

--------------------------------------------------------------------------------
*(Petitioner and affiliation, if any)*

By _____      _____
     *(Signature of representative or person filing petition)*        *(Title, if any)*

Address _____      _____
       *(Street and number, city, State, and ZIP Code)*        *(Telephone number)*

WILLFULLY FALSE STATEMENT ON THIS PETITION CAN BE PUNISHED BY FINE AND IMPRISONMENT (U.S. CODE, TITLE 18, SECTION 1001)

GPO 896-179

# Activity 5-36: Collective Bargaining Worksheet     Chapter 22

Name _____     Score _____

Date _____     Period _____

Complete the collective bargaining worksheet below.

| NEGOTIATING TEAMS | |
| --- | --- |
| **MANAGEMENT** | **UNION** |
| Chief negotiator: _____ | Chief negotiator: _____ |
| Team members: _____ | Team members: _____ |
| _____ | _____ |
| _____ | _____ |

| NEGOTIATING POSITIONS | |
| --- | --- |
| **MANAGEMENT** | **UNION** |
| 1. Rights of management: | |
| 2. Union security: | |
| 3. Wages: | |

*(Continued)*

Name _____

| MANAGEMENT | UNION |
|---|---|
| 4. Hours: | |
| 5. Benefits: | |
| 6. Working conditions: | |
| 7. Grievance procedure: | |
| 8. Employment and dismissal policy: | |

# Activity 5-37: Employee Grievance Sheet

**Chapter 22**

Name _____

Date _____

Score _____

Period _____

Complete the employee grievance sheet below.

| CONTRACT PROVISION INVOLVES |
|---|
| _____<br>_____ |

| NATURE OF GRIEVANCE |
|---|
| <br><br><br><br>Employee: _____  Steward: _____  Date: _____ |

| STEP 1: SUPERVISOR'S ANSWER: |
|---|
| <br><br><br>Signed:_____  Date: _____  Settled? ☐ Yes  ☐ No |

| STEP 2: DEPARTMENT MANAGER'S ANSWER: |
|---|
| <br><br><br>Signed:_____  Date: _____  Settled? ☐ Yes  ☐ No |

| STEP 3: LABOR RELATIONS MANAGER'S ANSWER: |
|---|
| <br><br><br>Signed:_____  Date: _____  Settled? ☐ Yes  ☐ No |

**Activity 5-38: Unions**

Name _____

Date _____

**Chapter 22**

Score _____

Period _____

You have now completed several days in your student-run enterprise. Imagine that you plan to make your living in such a factory.

1.  Have you seen anything that would make the workers want to form a union?
    _____ Yes.   _____ No.

2.  If you answered "yes," list some of the things which cause the workers to want a union:

    _____

    _____

    _____

    _____

    _____

3.  Would the union probably be an industrial or a craft union?
    _____ Industrial.   _____ Craft.

4.  If you were to form a union, what benefits would you try to gain for the workers?

    a.  _____

    b.  _____

    c.  _____

    d.  _____

    e.  _____

    f.  _____

5.  Would you want your union to be affiliated with the AFL-CIO?
    _____ Yes.   _____ No.

    Why? _____

    _____

    _____

    _____

    _____

## Activity 5-39: Company Identity Program    Chapter 23

Name _____

Date _____

Score _____

Period _____

Complete the following exercise.

| COMPANY NAMES |
|---|

Suggest three possible names for the company:

    1. _____

    2. _____

    3. _____

| ADVERTISING SLOGANS |
|---|

Suggest three advertising slogans for the product or company:

    1. _____

    2. _____

    3. _____

| COMPANY TRADEMARKS OR LOGOS |
|---|

Sketch two suggested logos for the company on the grids below.

## Activity 5-40: Package Design Sheet     Chapter 23

Name _____

Date _____

Score _____

Period _____

Product name: _____    Prepared by: _____

Design a package by drawing the outline of the package (allowing for all folds, cuts, and windows), dimensioning all features, and preparing a graphic layout for the type and artwork that will be printed on the package. NOTE: A tracing paper overlay may be used for the graphics.

## Activity 5-41: Advertising Layout Sheet    Chapter 23

Name _____    Score _____

Date _____    Period _____

Complete this exercise as you plan the advertising layout of your product.

Product: _____    Layout prepared by: _____

Theme: _____

Message: _____

_____

_____

_____

_____

_____

_____

_____

_____

_____

_____

_____

_____

_____

_____

_____

_____

_____

_____

_____

_____

*(Continued)*

Prepare a layout for the advertising copy on the previous page. Indicate type sizes and styles, colors, and the type and location of all graphics.

## Activity 5-42: Market Survey

Name _____

Date _____

Score _____

Period _____

Reproduce the survey below and ask potential customers to complete it.

---

### MARKET SURVEY

Complete the following form for the selected product.

PERSONAL INFORMATION:

1. Age:

   5 — 10

   10 — 15

   15 — 20

   Over 20

2. Sex:

   Male

   Female

3. Marital status:

   Married

   Single

4. Education (highest level completed):

   Grade School

   High School

   College

5. Place of residence:

   House

   Mobile home

   Duplex

   Apartment

MARKET INFORMATION:

6. Have you seen a similar product before?

   Yes          No

7. If yes, where?

   In a store

   On television

   At a friend's house

   Other (specify)

   _____

8. Do you own a similar product?

   Yes          No

9. If so, how long have your owned it?

   Less than a year

   1 — 2 years

   2 — 5 years

   Over 5 years

10. Would you buy this product if it were available?

    Yes

    No

11. If yes, how much would you pay for it?

    $0.50 — $1.00

    $1.00 — $1.50

    $1.50 — $2.00

    $2.00 — $2.50

    $2.50 — $3.00

    Over $3.00

PRODUCT INFORMATION:

12. Would you like the product to be:

    A light color

    A dark color

13. Would you like this product to be:

    The size it is

    Larger

    Smaller

14. What changes would you like to see made in the product design? (Specify.) _____

    _____

---

# Activity 5-43: Market Survey Data Results     Chapter 23

Name _____     Score _____

Date _____     Period _____

Using data obtained from the market survey on the previous page, compile the survey results below.

**Product:** _____

**Market Data:**
The recommended product has the following market appeal:

1. _____ percent of those surveyed would purchase at least one product. They suggested a selling price from $_____ to _____ with an average price of $_____ per unit.

2. _____ percent of those surveyed were males and _____ percent were females. _____ percent of the males and _____ percent of the females surveyed indicated they were interested in buying the product.

3. The age group surveyed ranged from _____ years to _____ years old. The following data by age group was collected:

| Age range | Would buy | Would not buy |
|-----------|-----------|---------------|
|           |           |               |
|           |           |               |
|           |           |               |
|           |           |               |
|           |           |               |
|           |           |               |
|           |           |               |
|           |           |               |

4. From the survey results it was found that the most likely market for the product is _____ from the age of _____ to _____ and the best selling
   (male or female)
   price seems to be $_____ per unit.

**Product data:**
The market reacted to the recommended product in the following ways:

1. _____ percent of the market liked the product as it was designed. The remainder recommended the following changes:

   _____ percentage suggested _____

   _____ percentage suggested _____

   _____ percentage suggested _____

2. _____ percent of the market thought the product should be _____
   (color, wood, etc.)

   while _____ indicated a preference of _____. Other responses included choices

   for _____, _____, and _____.

*(Continued)*

Name _____

## Competitive data:

A study of the competition found the following products which seem to compete for sales with the recommended product:

| Product | Where sold | Price |
|---|---|---|
| | | |
| | | |
| | | |
| | | |
| | | |
| | | |
| | | |
| | | |
| | | |
| | | |
| | | |

## Financial data:

The market and cost studies provided the following financial data:

Selling price:_____ . . . . . . . . . . . .                    $_____

Product cost:

    Manufacturing cost . . . . . . . . . _____

    Marketing cost . . . . . . . . . . . . _____

    Administrative cost . . . . . . . . . _____

Total cost . . . . . . . . . . . . . . . . . . . . . . . . .                    $_____

Profit . . . . . . . . . . . . . . . . . . . . . . . . . . . . .                    $_____

## Recommendation:

In review of the above data, the product design team recommends that the company (build) (not build) the product under consideration.

Signed: _____ Design Team Leader

# Activity 5-44: Sales Order Forms

## Chapter 23

Name _____

Date _____

Score _____

Period _____

NOTE: Photocopy these forms on three colors of paper (white, pink, and yellow) and collate them into packages of three. White is the customer's copy, yellow is the salesperson's, and pink is turned into the sales office.

---

(Overlay company name here.)

Customer's name: _____     Order number: _____

Address: _____

_____

| Quantity | Item | Unit price | Total price |
|----------|------|------------|-------------|
|  |  | $ | $ |
|  |  | $ | $ |

Salesperson: _____     Total amount of sale: $_____

---

(Overlay company name here.)

Customer's name: _____     Order number: _____

Address: _____

_____

| Quantity | Item | Unit price | Total price |
|----------|------|------------|-------------|
|  |  | $ | $ |
|  |  | $ | $ |

Salesperson: _____     Total amount of sale: $_____

---

(Overlay company name here.)

Customer's name: _____     Order number: _____

Address: _____

_____

| Quantity | Item | Unit price | Total price |
|----------|------|------------|-------------|
|  |  | $ | $ |
|  |  | $ | $ |

Salesperson: _____     Total amount of sale: $_____

# Activity 5-45: Sales Record Sheet

## Chapter 23

Name _____

Date _____

Score _____

Period _____

Complete the sales record sheet below.

Company name: _____ Date:_____

For the week from: _____ to: _____ Prepared by: _____

| Salesperson's name | Sales to date (Start) | Sales reported | | | | | Total to date (End) |
|---|---|---|---|---|---|---|---|
| | | Mon. | Tue. | Wed. | Thur. | Fri. | |
| | | | | | | | |
| | | | | | | | |
| | | | | | | | |
| | | | | | | | |
| | | | | | | | |
| | | | | | | | |
| | | | | | | | |
| | | | | | | | |
| | | | | | | | |
| | | | | | | | |
| | | | | | | | |
| | | | | | | | |
| | | | | | | | |
| | | | | | | | |
| | | | | | | | |
| | | | | | | | |
| | | | | | | | |
| | | | | | | | |
| | | | | | | | |
| | | | | | | | |
| | | | | | | | |
| | | | | | | | |

# Activity 5-46: Finished Goods Inventory

## Chapter 23

Name _____

Score _____

Date _____

Period _____

Complete the finished goods inventory form below.

Product: _____

Recorded by: _____

| Date | Balance on hand | Received from production | Returned from sales/shipping | Shipped to customers | New balance on hand |
|------|----------------|--------------------------|------------------------------|----------------------|---------------------|
|      |                |                          |                              |                      |                     |
|      |                |                          |                              |                      |                     |
|      |                |                          |                              |                      |                     |
|      |                |                          |                              |                      |                     |
|      |                |                          |                              |                      |                     |
|      |                |                          |                              |                      |                     |
|      |                |                          |                              |                      |                     |
|      |                |                          |                              |                      |                     |
|      |                |                          |                              |                      |                     |
|      |                |                          |                              |                      |                     |
|      |                |                          |                              |                      |                     |
|      |                |                          |                              |                      |                     |
|      |                |                          |                              |                      |                     |
|      |                |                          |                              |                      |                     |
|      |                |                          |                              |                      |                     |
|      |                |                          |                              |                      |                     |
|      |                |                          |                              |                      |                     |
|      |                |                          |                              |                      |                     |
|      |                |                          |                              |                      |                     |
|      |                |                          |                              |                      |                     |
|      |                |                          |                              |                      |                     |
|      |                |                          |                              |                      |                     |
|      |                |                          |                              |                      |                     |
|      |                |                          |                              |                      |                     |

# Activity 5-47: General Ledger

**Chapter 24**

Name _____

Score _____

Date _____

Period _____

Complete the general ledger below.

Company name: _____

Accounting period is from: _____ to _____ Prepared by: _____

| Date | Description | Budget code | Debit | | Credit | |
|------|-------------|-------------|-------|--|--------|--|
|      |             |             |       |  |        |  |
|      |             |             |       |  |        |  |
|      |             |             |       |  |        |  |
|      |             |             |       |  |        |  |
|      |             |             |       |  |        |  |
|      |             |             |       |  |        |  |
|      |             |             |       |  |        |  |
|      |             |             |       |  |        |  |
|      |             |             |       |  |        |  |
|      |             |             |       |  |        |  |
|      |             |             |       |  |        |  |
|      |             |             |       |  |        |  |
|      |             |             |       |  |        |  |
|      |             |             |       |  |        |  |
|      |             |             |       |  |        |  |
|      |             |             |       |  |        |  |
|      |             |             |       |  |        |  |
|      |             |             |       |  |        |  |
|      |             |             |       |  |        |  |
|      |             |             |       |  |        |  |
|      |             |             |       |  |        |  |
|      |             |             |       |  |        |  |

# Activity 5-48: Cost Accounting Ledger

# Chapter 24

Name _____

Score _____

Date _____

Period _____

Complete the cost accounting ledger below.

Department or profit center: _____

Accounting period is from: _____ to _____ Prepared by: _____

| Date | Description | Debit | Credit |
|------|-------------|-------|--------|
|      |             |       |        |
|      |             |       |        |
|      |             |       |        |
|      |             |       |        |
|      |             |       |        |
|      |             |       |        |
|      |             |       |        |
|      |             |       |        |
|      |             |       |        |
|      |             |       |        |
|      |             |       |        |
|      |             |       |        |
|      |             |       |        |
|      |             |       |        |
|      |             |       |        |
|      |             |       |        |
|      |             |       |        |
|      |             |       |        |
|      |             |       |        |
|      |             |       |        |
|      |             |       |        |
|      |             |       |        |
|      |             |       |        |

# Activity 5-49: Balance Sheet

## Chapter 24

Name _____

Date _____

Score _____

Period _____

Complete the balance sheet below.

Company name: _____

Prepared by: _____

| ASSETS | | |
|---|---|---|
| **CURRENT ASSETS** | | |
| Cash | $_____ | |
| Inventories | | |
| Finished products | $_____ | |
| Work-in-process | $_____ | |
| Raw materials | $_____ | |
| Other current assets | $_____ | |
| **TOTAL CURRENT ASSETS** | | $_____ |
| **FIXED ASSETS** | | |
| Production equipment | $_____ | |
| Less depreciation | $_____ | |
| **TOTAL FIXED ASSETS** | | $_____ |
| **TOTAL ASSETS** | | $_____ |

| LIABILITIES | | |
|---|---|---|
| **CURRENT LIABILITIES** | | |
| Notes (Working capital loans) | $_____ | |
| Accounts payable (Bills) | $_____ | |
| Taxes on income | $_____ | |
| Dividends payable | $_____ | |
| **TOTAL CURRENT LIABILITIES** | | $_____ |

| OWNER'S EQUITY | | |
|---|---|---|
| Common stock: _____ shares at _____ par value | | $_____ |
| Retained earnings | | $_____ |
| **TOTAL OWNER'S EQUITY** | | $_____ |
| **TOTAL LIABILITIES AND OWNER'S EQUITY** | | $_____ |

# Activity 5-50: Income Statement                    Chapter 24

Name _____          Score _____

Date _____          Period _____

Using the form below, prepare an income statement.

Company name: _____

Prepared by: _____

| REVENUE | | |
|---|---|---|
| Sale of finished products | $_____ | |
| Sale of raw material inventory | $_____ | |
| Sale of production equipment | $_____ | |
| **TOTAL REVENUE** | | $_____ |
| **COST OF GOODS SOLD** | | |
| Production costs (materials and labor) | $_____ | |
| Marketing costs | $_____ | |
| General administration costs | $_____ | |
| Interest and debt expenses | $_____ | |
| Other costs | $_____ | |
| **TOTAL COSTS AND EXPENSES** | | $_____ |
| **INCOME** | | |
| Taxes on income | $_____ | |
| Income before taxes | $_____ | |
| **NET INCOME** | | $_____ |
| **RETAINED EARNINGS** | | |
| Balance at beginning of year | $_____ | |
| Income from the year | $_____ | |
| **TOTAL** | | $_____ |
| Less dividends | $_____ | |
| **TOTAL LIABILITIES AND OWNER'S EQUITY** | | $_____ |

## Activity 5-51: Articles of Dissolution          Chapter 25

Name _____          Score _____

Date _____          Period _____

As you close your company, prepare the articles of dissolution below.

Prepared by: _____

| ARTICLES OF DISSOLUTION |
|---|
| The undersigned officers of _____ (Hereinafter referred to as the ''Corporation''), desiring to give notice of corporate action effectuating the dissolution of the Corporation. |

| ARTICLE 1 — Name |
|---|
| The name of the Corporation is_____ |

| ARTICLE 2 — Principal office |
|---|
| The Post Office address of the principal office of the Corporation is: |
| _____<br>(Number and Street or Building) |
| _____<br>(City, State and Zip Code) |

| ARTICLE 3 — Date and copy of notice of Shareholders' meeting |
|---|
| The date of the meeting of the Shareholders of the Corporation, called to consider the dissolution was _____, 19____; and a copy of the notice of such meeting is here set forth as follows: |

| ARTICLE 4 — Resolution of Shareholders |
|---|
| A copy of the resolution adopted at such meeting, or by unanimous written consent without a meeting, authorizing the dissolution, is here forth as follows: |

*(Continued)*

Name _____

---

**ARTICLE 5 — Manner of adoption and vote**

The Board of Directors of the Corporation, at a meeting thereof, duly called, constituted, and held on _____, 19____, at which a quorum was present, duly adopted a resolution submitting the question of dissolving the Corporation to a vote of the Shareholders at a meeting to be held on _____, 19____.

The Shareholders of the Corporation, entitled to vote in respect of such dissolution, adopted the resolution set forth in Article 4.

---

**ARTICLE 6 — Directors and officers**

**Section 1: Directors:** The names and post office addresses of the Board of Directors of the Corporation are:

| NAME | NUMBER AND STREET OR BUILDING | CITY | STATE | ZIP CODE |
|------|-------------------------------|------|-------|----------|
| | | | | |
| | | | | |
| | | | | |
| | | | | |
| | | | | |
| | | | | |

**Section 2: Officers:** The names and post office addresses of the officers of the Corporation are:

| NAME | NUMBER AND STREET OR BUILDING | CITY | STATE | ZIP CODE |
|------|-------------------------------|------|-------|----------|
| | | | | |
| | | | | |
| | | | | |
| | | | | |
| | | | | |

IN WITNESS WHEREOF, the undersigned officers execute these articles of Dissolution, and certify to the truth of the facts herein stated, this _____ day of _____, 19____.

_____     _____
(Written Signature)                          (Written Signature)
President of the Company                 Secretary of State

# Section 6
# AUTOMATING MANUFACTURING SYSTEMS

**Activity 6-1: Automation in Manufacturing**          **Chapter 26**

Name _____          Score _____

Date _____          Period _____

---

### THE CHALLENGE

Automation involves systems that automatically control and complete a task. Select one of the tasks below.

☐ Erase a chalk board.

☐ Make toast.

☐ Feed a dog at the same time each day.

☐ Bring the mail from a mail box.

☐ Take the garbage to the curb.

☐ Open a window when a room is too hot.

---

### THE TASK

List the actions that it would take to complete the selected task. For example, to turn off an alarm clock, the system would have to sense the correct sound, then turn off the clock. Steps to complete the task selected above:

1. _____

2. _____

3. _____

4. _____

5. _____

6. _____

7. _____

8. _____

9. _____

10. _____

11. _____

12. _____

*(Continued)*

Name _____

| THE SOLUTION |
|---|

Sketch your automated system below.

## Activity 6-2: Computers and Product Design    Chapter 27

Name _____    Score _____

Date _____    Period _____

Place a check next to the statement that applies to your situation and answer it in the space provided.

_____ If you used a computer in designing a product for your enterprise, describe the steps you used in the space below.

_____ If you did not use a computer in designing a product for your enterprise, describe how a computer may have made the design job more efficient.

_____

_____

_____

_____

_____

_____

_____

_____

Describe what you feel are the advantages of using computers in product design in relation to each of the following:

Engineering design: _____

_____

_____

_____

Design analysis: _____

_____

_____

_____

Design presentation: _____

_____

_____

_____

# Activity 6-3: Producing a Design on the Computer Chapter 27

Name _____     Score _____

Date _____     Period _____

---

## THE TASK

Use a CAD or graphics software program on a computer to prepare a three-dimensional view drawing of the product shown below.

---

## THE PRODUCT

**SPECIFICATIONS:**

Block is 4'' square and 1/2'' thick.
16 holes are 1/8'' diameter × 1/4'' deep.
Holes are 1/2'' from the edge of the block.
Holes are located 1'' on center.

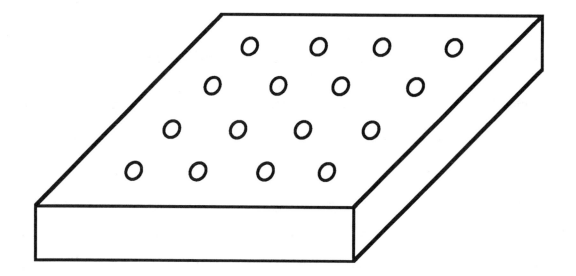

## Activity 6-4: Computers in Manufacturing

**Chapter 28**

Name _____

Date _____

Score _____

Period _____

Place a check next to the statement that applies to your situation and answer it in the space provided.

_____ If you used a computer in manufacturing a product for your enterprise, describe the steps you used in the space below.

_____ If you did not use a computer in manufacturing a product for your enterprise, describe how a computer may have made the manufacturing task more efficient.

_____

_____

_____

_____

_____

_____

_____

Describe what you feel are the advantages of using computers in relation to each of the following:

Material processing: _____

_____

_____

_____

_____

Management: _____

_____

_____

_____

_____

# Activity 6-5: Designing a Manufacturing Cell

## Chapter 28

Name _____

Date _____

Score _____

Period _____

| THE TASK |
|---|

Design a manufacturing cell to produce the product shown in Activity 6-3. Start the process with boards that are 1/2'' thick, 4'' wide, and 8' long.

| THE PRODUCT |
|---|

**SIZES:**

Block:
 1/2 " x 4" x 4"

Holes
 1/8" Dia. x 1/4" Deep

Cut to length
Drill holes
Sand faces
Sand edges
Sand ends
Inspect

| THE MANUFACTURING CELL |
|---|

Sketch the manufacturing cell in the space provided.

# Section 7
# MANUFACTURING, TECHNOLOGY, AND YOU

**Activity 7-1: Manufacturing and the Future**         **Chapter 29**

Name _____          Score _____

Date _____          Period _____

The future may see people building and operating manufacturing facilities in space. On the grid below, design a manufacturing space station. Label all sections. On the following page, describe the space station.

*(Continued)*

Name _____

Describe each of the following aspects of the manufacturing space station you designed:

Item that would be manufactured: _____

_____

_____

Number of people that could work in the station: _____

How manufacturing would take place: _____

_____

_____

_____

_____

_____

How power would be generated: _____

_____

_____

_____

_____

Transporting: _____

_____

_____

_____

_____

_____

Waste disposal: _____

_____

_____

Working conditions: _____

_____

_____

## Activity 7-2: Encouraging Teamwork

Name _____

Date _____

Score _____

Period _____

| THE CHALLENGE |
| --- |
| Select a task that you think a team of students from your school could complete. Then, on this form, describe how you would create an atmosphere so that the team would complete the task on time and with excellence.<br><br>Task: _____ |

| TEAM ATTITUDE |
| --- |
| |

| SHARED VISION |
| --- |
| |

| PLAN OF ACTION |
| --- |

1. _____
2. _____
3. _____
4. _____
5. _____
6. _____

7. _____
8. _____
9. _____
10. _____
11. _____
12. _____

# Activity 7-3: Balancing a Checkbook

## Chapter 30

Name _____

Date _____

Score _____

Period _____

Manufacturing knowledge can be applied to everyday, personal activities. Use your knowlege of basic accounting principles to balance the following personal checkbook. The account has the following activities for the month:

Starting balance: $286.41

| 6/2 | Deposit paycheck $376.21 |
| 6/4 | Check # 171 to Pete's Grocery $54.21 |
| 6/5 | Check # 172 to Bazaar Clothing $29.95 |
| 6/10 | Check # 173 to Mohawk Power Company $38.74 |
| 6/15 | Check # 174 to Auto Loan Company $224.50 |
| 6/15 | Check # 175 to Rent $285.00 |
| 6/16 | Deposit paycheck $376.21 |
| 6/20 | Check # 176 to Great Lakes Phone Co. $31.42 |
| 6/21 | Check # 177 to Pete's Grocery $62.71 |
| 6/26 | Check # 178 to Sharp Oil Company $37.94 |
| 6/29 | Check # 179 to Payless Discount Store $18.43 |

---

**CHECK LEDGER**

| Number | Date | Desription of transaction | Payment/Debit | | Deposit/Credit | | Balance | |
|--------|------|---------------------------|---------------|--|----------------|--|---------|--|
| | | **Starting balance** | | | | | | |
| | | | | | | | | |
| | | | | | | | | |
| | | | | | | | | |
| | | | | | | | | |
| | | | | | | | | |
| | | | | | | | | |
| | | | | | | | | |
| | | | | | | | | |
| | | | | | | | | |
| | | | | | | | | |
| | | | | | | | | |
| | | | | | | | | |
| | | | | | | | | |
| | | | | | | | | |
| | | | | | | | | |
| | | | | | | | | |
| | | | | | | | | |
| | | | | | | | | |

**Activity 7-4: Career Preparation**                **Chapter 30**

Name _____          Score _____

Date _____          Period _____

Answer the following questions about career preparation.

1. Describe each of the following:

   Your abilities: _____

   _____

   _____

   _____

   _____

   Your limitations: _____

   _____

   _____

   _____

   _____

   Your interests: _____

   _____

   _____

   _____

   _____

   The life-style you want to have: _____

   _____

   _____

   _____

   _____

2. At this point, what occupation or career would best fill your needs? (Consider your responses to number 1 above.)

   _____

3. What do you feel are suitable ''backup'' choices?

   _____

   _____

   _____

4. What ''exploration'' courses seem best suited to help ''firm up'' your career decision?

   Academic courses: _____

   _____

   Technology courses: _____

   _____

*(Continued)*

5. What additional knowledge do you need about yourself in order to make a better career decision?

_____

_____

_____

How do you plan to obtain this information?

_____

_____

_____

6. What additional knowledge do you need about the world of work in order to make a better career decision?

_____

_____

_____

How do you plan to obtain this needed information?

_____

_____

_____

7. What program (or combination of courses) would best help you obtain job skills related to your career choices?

_____

_____

_____

8. After graduation from high school, what steps should be taken to further prepare you for your present (but tentative) career goal?

_____

_____

_____

_____

_____

_____

**Activity 7-5: Career Investigation**   **Chapter 30**

Name _____   Score _____

Date _____   Period _____

Investigate a career of your choice in the manufacturing field. If possible, interview someone with that career. However, you may use library materials, information from your guidance counselor, etc. Find out the following information:

Career: _____

Education or training needed: _____

_____

_____

_____

_____

Salary range: _____

Description of duties: _____

_____

_____

_____

_____

_____

Hours: _____

Working conditions: _____

_____

_____

_____

Advantages of this career: _____

_____

_____

_____

_____

Disadvantages of this career: _____

_____

_____

_____

_____

Other comments: _____

_____

_____